Jupiter

The King of the Heavens

JD ARDEN

Preface: A Titan Among Planets

Jupiter stands as a colossus in the celestial ballet of our solar system, a planet so massive that its gravitational influence extends far beyond its orbit. With a mass over 300 times that of Earth and a diameter more than 11 times as large, Jupiter is not just the largest planet—it is the gravitational anchor of the solar system, shaping the paths of comets, asteroids, and even its planetary siblings.

From its iconic Great Red Spot to its retinue of more than 90 moons, Jupiter represents a realm of extremes. Its swirling bands of clouds hide mysteries of chemistry and dynamics that defy easy explanation. Its magnetosphere, the largest structure in the solar system, dwarfs even the Sun. Yet, for all its grandeur, Jupiter also exerts a quiet, stabilizing influence, acting as both a protector and a potential threat to the inner planets.

In mythology, Jupiter—or Zeus to the Greeks—was the king of the gods, wielding thunderbolts and presiding over the heavens. The planet named in his honor is no less majestic, a world of storms and power that embodies the creative and destructive forces of the cosmos. To study Jupiter is to confront the duality of nature itself: the balance between chaos and order, beauty and danger, stability and change.

This book explores Jupiter not only as a physical entity but also as a symbol of cosmic grandeur and complexity. From the storms that rage across its surface to the icy oceans of Europa and the volcanic infernos of Io, Jupiter challenges our understanding of what planets can be. It is a reminder that our solar system, for all its apparent familiarity, is filled with wonders beyond imagination.

Chapter 1: The Great Red Spot

Jupiter's Great Red Spot is a planetary enigma, a storm so vast that it could swallow Earth whole, and so enduring that it has raged for centuries. Its swirling, crimson-hued clouds, visible even through modest telescopes, make it one of the most recognizable features in the solar system. Yet, despite centuries of observation, the Great Red Spot remains a mystery—its origins, longevity, and future still elude definitive explanation.

The first recorded observation of the Great Red Spot dates back to the 17th century, when astronomers like Giovanni Cassini and Robert Hooke described a persistent, reddish oval on Jupiter's surface. While it is uncertain whether these early observations refer to the same storm visible today, the Spot's longevity is remarkable. Most storms on Earth dissipate within days or weeks, yet the Great Red Spot has persisted for at least 350 years, and possibly much longer.

At its peak, the Great Red Spot measured over 40,000 kilometers in diameter—large enough to fit three Earths side by side. In recent decades, however, the storm has been shrinking, its once-elliptical shape becoming more circular, and its diameter now reduced to about 16,000 kilometers. This contraction has puzzled scientists, leading to questions about whether the Great Red Spot is fading, or merely transitioning to a new phase.

The storm's striking red color is another source of intrigue. Its hue varies over time, ranging from deep brick red to pale salmon, suggesting changes in its chemical composition or the altitude of its clouds. Scientists hypothesize that the color results from the interaction of ultraviolet sunlight with compounds in Jupiter's atmosphere, such as ammonia and hydrocarbons, though the precise mechanisms remain debated.

The Great Red Spot is driven by the dynamics of Jupiter's atmosphere, a turbulent system dominated by alternating bands of eastward and westward winds known as jet streams. These jet streams, which reach speeds of hundreds of kilometers per hour, create the stark boundaries between Jupiter's light-colored zones and dark-colored belts. The Great

Jupiter

Red Spot sits between two jet streams, spinning counterclockwise in a perpetual state of high-pressure turbulence.

What sustains the Great Red Spot's longevity is a key question. On Earth, storms lose energy as they interact with landmasses or the ocean surface, but Jupiter lacks such solid surfaces. Instead, the storm may draw energy from the surrounding jet streams or the planet's internal heat, which is generated by the slow gravitational contraction of its immense mass. Jupiter radiates more energy into space than it receives from the Sun, and this heat may play a role in fueling its weather systems.

Despite its enduring presence, the Great Red Spot is not immutable. Observations have revealed fluctuations in its size, shape, and color over the centuries, and some scientists believe it may eventually dissipate entirely. If the Spot is indeed shrinking, it raises questions about the timescales of planetary weather and the forces that govern storms on gas giants. Could another storm rise to take its place, or is the Great Red Spot a unique phenomenon tied to a specific era in Jupiter's history?

Beyond its scientific significance, the Great Red Spot holds a deeper, almost philosophical allure. It is a reminder of the power and scale of natural forces, a storm that has endured through centuries of human history, from the Age of Enlightenment to the modern Space Age. It is both a symbol of chaos and a testament to the order that governs the universe—a storm that rages endlessly yet follows the principles of physics.

The Great Red Spot also challenges our understanding of planetary atmospheres. As we study it, we gain insights into the dynamics of weather not just on Jupiter, but on other gas giants, exoplanets, and even Earth itself. Its persistence and complexity highlight the interconnectedness of planetary systems, where heat, rotation, and chemistry combine to create phenomena far beyond human experience.

Standing on the edge of the Great Red Spot—if such a feat were possible— would be an experience like no other. The winds would howl at speeds of hundreds of kilometers per hour, the ammonia-rich clouds swirling in a kaleidoscope of reds, yellows, and whites. Above, the sky would be a deep, alien blue, transitioning to the blackness of space. It is a vision

Jupiter

both terrifying and sublime, a reminder of the vastness and power of the cosmos.

The Great Red Spot is more than a storm; it is a monument to Jupiter's grandeur, a feature that has inspired awe and wonder for generations. Its mysteries, though profound, invite us to learn more—not only about Jupiter, but about the forces that shape planets and weather systems across the universe. In its swirling depths, the Great Red Spot holds the story of a planet that defies comparison, a world of extremes that continues to challenge our understanding of what is possible.

Chapter 2: A Gravitational Shield

Jupiter's immense size and powerful gravity make it more than just a gas giant—it is a protector, a sentinel that has profoundly shaped the solar system's history and continues to safeguard its inner planets, including Earth. Dubbed a "gravitational shield," Jupiter acts as a cosmic vacuum cleaner, pulling in or deflecting comets, asteroids, and other debris that might otherwise pose a threat to the inner planets. This role, while protective, is also complex, as Jupiter's influence can sometimes redirect objects toward Earth instead of away from it.

Jupiter's gravitational dominance is unmatched in the solar system. With a mass more than twice that of all the other planets combined, its gravitational pull extends far and wide, sculpting the orbits of countless objects in the asteroid belt, the Kuiper Belt, and even the distant Oort Cloud. For billions of years, this gravitational reach has acted as both a guardian and a disruptor, shaping the architecture of the solar system and influencing the fate of countless celestial bodies.

One of Jupiter's most significant roles is its ability to capture and neutralize comets and asteroids. Many of these objects, originating in the outer solar system, are drawn inward by the Sun's gravity. Without Jupiter's intervention, some of these objects might cross Earth's orbit, posing a threat of catastrophic impacts. Instead, Jupiter often captures these wanderers, either absorbing them into its own atmosphere or flinging them into distant orbits where they are no longer a danger to the inner planets.

A dramatic example of Jupiter's protective role occurred in 1994, when the comet **Shoemaker-Levy 9** collided with the planet. The comet, broken into fragments by Jupiter's tidal forces, left a series of dark scars across the gas giant's atmosphere as each fragment impacted. These scars, visible even through amateur telescopes, were a stark reminder of Jupiter's role as a cosmic shield. Had Shoemaker-Levy 9 bypassed Jupiter and struck Earth instead, the consequences would have been devastating.

Jupiter's gravitational influence also shapes the asteroid belt, a region of rocky debris located between Mars and Jupiter. Without Jupiter's presence, the material in the asteroid belt might have coalesced into a planet. Instead, Jupiter's gravity disrupts the accretion process, preventing the formation of a significant body and maintaining the asteroid belt as a collection of smaller objects. By confining these objects to stable orbits, Jupiter reduces the likelihood of them being ejected toward the inner planets.

However, Jupiter's role is not purely protective. Its immense gravity can also destabilize objects, sending them hurtling inward toward Earth. Some of the asteroids and comets that cross Earth's orbit, known as **Near-Earth Objects (NEOs)**, owe their trajectories to gravitational interactions with Jupiter. In this sense, Jupiter is both a savior and a potential threat—a duality that highlights the complexity of its influence.

Jupiter's role as a gravitational shield extends beyond immediate protection. Its interactions with celestial bodies over billions of years have shaped the conditions for life on Earth. By clearing the early solar system of excess debris, Jupiter may have reduced the frequency of catastrophic impacts, allowing Earth's biosphere to develop relatively undisturbed. At the same time, periodic impacts—some influenced by Jupiter—may have delivered water and organic materials to Earth, seeding the conditions for life.

The so-called **Late Heavy Bombardment**, a period of intense asteroid and comet impacts approximately 4 billion years ago, may have been influenced by Jupiter's migration through the early solar system. During this time, gravitational interactions between Jupiter, Saturn, and the outer planets destabilized the orbits of smaller bodies, sending a barrage of impacts toward the inner planets. While destructive, these impacts likely played a role in shaping Earth's early environment, including the delivery of water and essential compounds.

Jupiter's gravitational reach also extends to distant realms of the solar system, influencing the Kuiper Belt and the Oort Cloud. These regions, home to icy bodies and long-period comets, are shaped by Jupiter's interactions with other planets and the Sun. By perturbing the orbits of objects in these regions, Jupiter periodically sends comets on long journeys through the solar system, where they become targets for scientific study and reminders of the dynamic nature of planetary systems.

Jupiter

Philosophically, Jupiter's role as a gravitational shield invites reflection on the balance between protection and risk. Its massive presence is both a stabilizing force and a source of potential chaos, a reminder that even the most beneficial influences carry inherent complexities. Jupiter is a guardian, but it is not infallible, and its interactions with the solar system are a testament to the unpredictable interplay of gravity, motion, and chance.

Jupiter's influence also extends beyond its physical interactions. Its role in shaping the solar system inspires awe, a reminder of the interconnectedness of planetary systems and the forces that govern them. Jupiter's gravitational dominance is a window into the processes that have shaped not only our solar system but also exoplanetary systems across the galaxy. By studying Jupiter, we gain insights into the formation and evolution of other planetary systems, deepening our understanding of the universe.

The idea of Jupiter as a protector is a powerful metaphor. It underscores the fragility of life on Earth, a world shielded from chaos by the gravitational forces of a distant giant. It also highlights the paradox of existence—a planet that protects and endangers, a force of stability and disruption. Jupiter is a reminder that life exists within a delicate balance, shaped by forces far beyond our control.

As we continue to study Jupiter, its role as a gravitational shield will remain a central theme, not only in understanding the solar system's history but also in appreciating the cosmic dynamics that make life on Earth possible. Jupiter, the king of the heavens, is a testament to the interplay of chaos and order, a force that protects and challenges, and a celestial guardian that stands watch over its planetary siblings.

Chapter 3: The Jovian Moons

Orbiting the immense bulk of Jupiter are more than 90 moons, a miniature solar system in their own right. Among these, four stand out as the most remarkable: **Io**, **Europa**, **Ganymede**, and **Callisto**—collectively known as the Galilean moons, named after Galileo Galilei, who first observed them in 1610. These moons are worlds of staggering diversity and intrigue, each offering unique insights into the processes that shape planetary systems and the potential for life beyond Earth.

The Galilean moons, though bound to Jupiter by gravity, are far from passive satellites. Each has its own character, shaped by the interplay of tidal forces, internal geology, and interactions with Jupiter's powerful magnetosphere. Together, they form a dynamic system that challenges our understanding of planetary formation, habitability, and the limits of what a moon can be.

Io: The Volcanic Inferno

Io, the closest of the Galilean moons to Jupiter, is a world of fire and sulfur. It is the most volcanically active body in the solar system, its surface dotted with hundreds of active volcanoes that spew molten lava and sulfurous gases into space. The cause of this relentless activity is **tidal heating**—the result of gravitational interactions between Io, Jupiter, and the other Galilean moons.

As Io orbits Jupiter, the immense gravitational pull of the planet and the rhythmic tugs from Europa and Ganymede stretch and compress its interior, generating frictional heat. This heat melts the moon's rock, driving the volcanic eruptions that continuously reshape its surface. Unlike Earth's volcanoes, which are confined to specific regions, Io's eruptions are global, creating a constantly changing landscape of lava flows, calderas, and sulfur plains.

Io's intense activity makes it a hostile world, its surface temperatures fluctuating wildly between its molten hot spots and its frigid plains. The moon's thin atmosphere, composed mainly of sulfur dioxide, offers no protection from Jupiter's radiation, and its proximity to the planet places it deep within Jupiter's magnetosphere. This environment bombards Io with charged particles, creating auroras and adding to the complexity of its already dynamic system.

Despite its hostility, Io offers a glimpse into the extremes of planetary geology. It demonstrates the power of tidal forces and provides a natural laboratory for studying the interactions between a moon and its parent planet.

Europa: The Ocean World

Europa, the second Galilean moon, could not be more different from Io. Its icy surface, crisscrossed with cracks and ridges, hints at a vast, liquid ocean hidden beneath. Europa's smooth, reflective exterior suggests a relatively young surface, constantly reshaped by processes linked to its subsurface ocean.

Scientists believe that Europa's ocean, kept liquid by tidal heating, could contain more water than all of Earth's oceans combined. This ocean, hidden beneath kilometers of ice, is of profound interest because it represents one of the most promising places in the solar system to search for life.

Europa's potential habitability stems from the interaction between its ocean and its rocky core. As tidal forces flex Europa's interior, they could drive hydrothermal activity on the ocean floor, similar to the deep-sea vents on Earth that teem with life. These vents, powered by heat and chemistry rather than sunlight, could provide the energy and nutrients needed to sustain microbial ecosystems.

Europa's icy surface also plays a role in its potential habitability. The cracks and plumes observed on its surface may act as conduits, allowing materials from the ocean to reach the surface and vice versa. These exchanges could create a dynamic system where organic molecules, energy, and nutrients cycle between the ocean and the ice shell.

The possibility of life on Europa has inspired numerous missions, including NASA's upcoming **Europa Clipper**, which will conduct detailed reconnaissance of the moon's surface, ice shell, and subsurface ocean. Europa's mysteries challenge our understanding of habitability, expanding the search for life beyond the traditional parameters of liquid water and sunlight.

Jupiter

Ganymede: The Giant Moon

Ganymede, the largest moon in the solar system, is a world of superlatives. Larger than the planet Mercury, it is the only moon known to have a magnetic field, a feature that adds complexity to its interactions with Jupiter's magnetosphere. Ganymede's surface is a mix of ancient, heavily cratered terrain and younger, grooved regions that hint at tectonic activity in the past.

Beneath its icy surface, Ganymede is thought to harbor a subsurface ocean, similar to Europa's, though buried deeper beneath its crust. This ocean may interact with the moon's rocky mantle, creating conditions that could support life. Ganymede's magnetic field, generated by a liquid iron or iron-sulfide core, offers additional protection from radiation, enhancing its potential habitability.

Ganymede's size and complexity make it a focal point for understanding the processes that shape icy moons. Its layered structure—a metal core, rocky mantle, and icy shell—resembles a miniature planet, offering clues about planetary differentiation and the evolution of moons around gas giants.

Callisto: The Ancient Cratered World

Callisto, the outermost of the Galilean moons, is a stark contrast to its siblings. Its surface is one of the oldest in the solar system, heavily cratered and largely unchanged for billions of years. Unlike Io, Europa, or Ganymede, Callisto shows little evidence of internal activity, its geology shaped primarily by impacts rather than tectonics or volcanism.

Despite its apparent dormancy, Callisto holds intrigue. Beneath its ancient surface, it may also harbor a subsurface ocean, though less active than Europa's or Ganymede's. This ocean, if it exists, would be shielded from radiation by Callisto's distance from Jupiter, making it a stable and quiet environment.

Callisto's isolation and stability make it an appealing target for future human exploration. Its low radiation levels and abundance of water ice could support habitats or serve as a resource base for missions to the outer solar system.

Jupiter

The Galilean moons are a testament to the diversity of planetary systems. Each offers a unique perspective on the processes that shape moons, planets, and the potential for life. Together, they challenge our understanding of what it means to be habitable, expanding the boundaries of where we might find life and how we explore the cosmos.

Jupiter's moons are not merely satellites; they are worlds unto themselves, each contributing to the grand story of the solar system. From the volcanic fury of Io to the icy oceans of Europa, the magnetic mysteries of Ganymede, and the ancient tranquility of Callisto, they are a reminder of the richness and complexity of the universe.

Chapter 4: Rings of the Giant

Jupiter, with its immense mass and swirling atmosphere, is often defined by its grandeur and dynamism. But hidden within this spectacle is a feature far more subtle: its rings. Unlike the brilliant, icy bands of Saturn, Jupiter's rings are faint and delicate, composed primarily of fine dust. They are almost invisible in visible light, detectable only through the keen eyes of spacecraft and infrared observations. Yet, these ethereal rings hold a quiet significance, offering insights into the processes that shape planetary systems and the interplay of moons, gravity, and time.

The discovery of Jupiter's rings came relatively late in our exploration of the solar system. It wasn't until 1979, when NASA's Voyager 1 spacecraft passed by the giant planet, that their existence was confirmed. Voyager's images revealed a faint, tenuous structure encircling Jupiter, a discovery that challenged assumptions about gas giants and their features. Subsequent missions, including Galileo and Juno, provided further details, painting a picture of a ring system that, while subtle, is rich in complexity.

Jupiter's rings are composed of three main components: a broad, faint halo closest to the planet, a brighter and thinner main ring, and an even fainter outer gossamer ring. These components are formed and maintained through the interactions of Jupiter's gravity, its moons, and the relentless bombardment of micrometeoroids. Unlike Saturn's rings, which are primarily made of ice, Jupiter's rings are composed of fine dust particles, originating from the surfaces of its inner moons.

The main ring, the brightest of the three, owes its existence to the moons Metis and Adrastea, two small, irregular satellites that orbit within the ring's bounds. These moons are continuously struck by meteoroids, creating a steady supply of dust that is pulled into the ring by Jupiter's gravity. The halo, which extends closer to the planet, forms as some of this dust spirals inward, influenced by Jupiter's intense magnetic field.

The gossamer ring, faint and diffuse, stretches farther outward and is associated with two additional moons, Amalthea and Thebe. Like their counterparts in the main ring, these moons contribute material through

Jupiter

the constant impacts of micrometeoroids, creating a tenuous structure that is almost ghostly in its delicacy.

What makes Jupiter's rings fascinating is their dynamism. Unlike the more stable icy rings of Saturn, Jupiter's rings are transient, their particles constantly replenished and lost over time. The interplay between the planet's gravity, the orbits of its moons, and the impacts that generate the dust creates a system in perpetual flux. This fragility makes Jupiter's rings a snapshot of an ongoing process, a reminder that even the most seemingly static features of the cosmos are subject to change.

Jupiter's rings also provide a window into the forces that shape planetary systems. Their composition and behavior are a testament to the interplay of gravity, collisions, and magnetism, forces that have sculpted the solar system for billions of years. By studying Jupiter's rings, scientists gain insights into the processes that govern not only the outer planets but also other ringed worlds and even the formation of planetary systems around distant stars.

The rings also hint at the history of Jupiter and its moons. The dust that makes up the rings carries the imprint of countless collisions, a record of the solar system's ceaseless activity. Each particle is a fragment of a larger story, one that spans the lifetimes of moons, meteoroids, and the forces that bind them.

In their quiet subtlety, Jupiter's rings challenge our expectations. They lack the grandeur of Saturn's icy bands or the dramatic arcs of Neptune's rings, but they possess a beauty all their own—a fragile, intricate structure that reflects the complexity of the giant planet they encircle. Jupiter's rings are a reminder that even the smallest, most understated features of the solar system hold significance, connecting moons, planets, and processes in ways that reveal the interconnectedness of the cosmos.

Chapter 5: Mysteries of the Magnetic Field

Jupiter's magnetic field is one of its most extraordinary and enigmatic features. Vast, powerful, and unlike anything found elsewhere in the solar system, it is a force that shapes the environment around the planet, creating a magnetosphere so immense that it could encompass the Sun and still leave room for Earth's orbit. This magnetic field is not just a feature of the planet; it is an active participant in the dynamics of the Jovian system, influencing its atmosphere, moons, and even the space around it. Understanding Jupiter's magnetic field is a journey into the heart of a phenomenon that defies earthly comparisons.

Generated by the movement of conductive material within its interior, Jupiter's magnetic field dwarfs Earth's in both size and strength. At its surface, the field is about 20,000 times stronger than Earth's, and it extends outward for millions of kilometers, forming a magnetosphere that stretches beyond the orbit of Saturn in some directions. This immense structure is not symmetrical; it is shaped and distorted by the solar wind, a stream of charged particles flowing outward from the Sun. On the side facing the Sun, the magnetosphere is compressed, while on the opposite side, it stretches into a vast tail that trails behind the planet.

The source of Jupiter's magnetic field lies deep within its interior, in a region of metallic hydrogen created by the planet's immense pressure. Unlike Earth, where the magnetic field is generated by the motion of molten iron in its core, Jupiter's field arises from the movement of this exotic form of hydrogen, which behaves like a liquid metal under the extreme conditions found within the planet. The rapid rotation of Jupiter, which completes a day in just under 10 hours, adds to the dynamo effect, amplifying the field's strength and complexity.

Jupiter's magnetic field is far from uniform. Observations by spacecraft like Juno have revealed a field with unexpected asymmetries and localized regions of intensity. Near the equator, a strange patch of magnetic strength, nicknamed the "Great Blue Spot," defies traditional models of planetary magnetism. These anomalies hint at dynamic processes occurring within the planet's interior, processes that remain poorly understood but offer tantalizing clues about how magnetic fields are generated and sustained.

One of the most dramatic consequences of Jupiter's magnetic field is its interaction with its moons, particularly Io. As Io orbits within Jupiter's magnetosphere, it acts as both a source and a conduit for electrical currents. Volcanic eruptions on Io spew vast quantities of sulfur dioxide gas, which becomes ionized and forms a plasma torus—a doughnut-shaped ring of charged particles that encircles Jupiter. This plasma is swept up by Jupiter's rotation, creating intense electrical currents that connect Io to the planet's poles, generating auroras far more powerful than those seen on Earth.

Jupiter's auroras, visible in ultraviolet and infrared light, are among the most energetic phenomena in the solar system. Unlike Earth's auroras, which are primarily driven by interactions between the solar wind and the magnetic field, Jupiter's auroras are largely powered by the planet's own rotation and the volcanic activity of Io. These displays are not just visual spectacles; they are windows into the complex interactions between Jupiter's magnetic field, its moons, and its atmosphere.

The magnetosphere also serves as a protective shield, deflecting the solar wind and creating a bubble of space dominated by Jupiter's magnetic influence. Within this bubble, charged particles are trapped, forming intense radiation belts that far exceed the levels found in Earth's Van Allen belts. These belts are a hazard to spacecraft and a challenge for future missions to Jupiter and its moons, requiring advanced shielding and careful navigation.

The immense power of Jupiter's magnetic field raises broader questions about the nature of magnetism in planetary systems. By studying Jupiter, scientists hope to better understand the dynamo processes that generate magnetic fields, not only in planets but also in stars and other celestial bodies. The insights gained from Jupiter could inform our understanding of Earth's own magnetic field, its role in protecting the atmosphere, and its interactions with the solar wind.

Jupiter's magnetic field also has implications for the search for life. Europa and Ganymede, two of Jupiter's largest moons, are thought to harbor subsurface oceans, and their interactions with the magnetic field could provide energy for potential ecosystems. The flux of charged particles and the induced electric currents within these oceans may create environments capable of supporting life, expanding the boundaries of habitability beyond traditional definitions.

Philosophically, Jupiter's magnetic field is a reminder of the unseen forces that shape the cosmos. It is a phenomenon of staggering scale, influencing not just the planet itself but the space far beyond. Its invisible lines of force connect moons, plasma, and particles in a complex dance, creating a system that is as dynamic as it is vast.

To stand within Jupiter's magnetosphere—if such a thing were possible—would be to experience an environment utterly alien to anything on Earth. The radiation would be lethal, the electric currents overwhelming, and the auroras blinding in their intensity. It is a realm that challenges our imagination, a testament to the power of natural forces that operate on a scale far beyond human experience.

Jupiter's magnetic field is not just a feature of the planet; it is a living system, constantly changing and interacting with its surroundings. It is a source of wonder and a subject of study, a phenomenon that invites us to look deeper into the processes that govern the universe. In its vastness and complexity, it reflects the grandeur of Jupiter itself—a giant among planets, a king of the heavens, and a reminder of the forces that shape our world and beyond.

Chapter 6: The Gas Giant's Atmosphere

Jupiter's atmosphere is an extraordinary tapestry of dynamic forces, layered clouds, and violent storms that defy comparison to anything on Earth. Stretching thousands of kilometers deep, it is a complex system dominated by hydrogen and helium, yet punctuated by traces of compounds like ammonia, methane, and water vapor that shape its appearance and behavior. Jupiter's atmosphere is not merely a shell surrounding the planet—it is a chaotic and ever-changing realm that encapsulates the planet's grandeur and complexity.

The atmosphere's most striking feature is its **banded structure**, a series of alternating light-colored zones and dark-colored belts that encircle the planet. These bands are the result of powerful jet streams that blow in opposite directions, reaching speeds of hundreds of kilometers per hour. The contrast between the zones and belts arises from differences in composition, temperature, and altitude, creating a visual effect that makes Jupiter instantly recognizable even through a small telescope.

The colors of Jupiter's atmosphere, ranging from creamy whites to ochre and deep reds, are shaped by the chemistry of its upper layers. The exact compounds responsible for these hues remain a subject of scientific inquiry, but they likely include complex organic molecules, sulfur compounds, and other trace elements. Ultraviolet radiation from the Sun interacts with these substances, producing the vibrant colors that ripple across the planet's surface.

Beneath the visible clouds lies a world of increasing pressure and temperature. The uppermost layer of clouds is composed primarily of ammonia ice crystals, forming a thin veil over deeper layers of ammonium hydrosulfide and water clouds. As one descends, the pressure increases dramatically, compressing the gases into a dense, hot mixture. While the visible atmosphere extends only a few hundred kilometers, the full depth of Jupiter's gaseous envelope is measured in thousands of kilometers, blending seamlessly into the liquid hydrogen that constitutes much of the planet's bulk.

Jupiter

The movement within Jupiter's atmosphere is dominated by immense storms and vortices, the most famous of which is the **Great Red Spot**. However, this iconic storm is only one of countless weather phenomena that mark Jupiter's surface. Smaller storms, often organized into cyclonic and anticyclonic patterns, dot the planet's bands, merging, dissipating, and reforming in an ongoing dance of atmospheric turbulence. These storms can last for years or even centuries, driven by the planet's rapid rotation and the internal heat that radiates outward from its core.

Unlike Earth, where weather is driven primarily by solar energy, Jupiter's atmosphere is powered largely by heat from the planet itself. Jupiter emits more energy than it receives from the Sun, a result of the slow contraction of its immense mass. This internal heat creates convection currents that rise and fall within the atmosphere, driving the formation of storms, clouds, and jet streams. The interplay of this internal energy with the planet's rotation and chemistry creates a system of unparalleled complexity.

Another hallmark of Jupiter's atmosphere is its **lightning**, some of the most powerful in the solar system. Detected by spacecraft and even visible from Earth-based telescopes, these strikes illuminate the cloud tops in brief, brilliant flashes. Jupiter's lightning is thought to occur in the deeper water clouds, where rising and falling air masses generate the necessary electrical charges. These storms, though similar in principle to those on Earth, dwarf anything found on our planet, with flashes that can be thousands of times more energetic.

The sheer scale of Jupiter's atmosphere is difficult to grasp. Its dynamic weather systems span thousands of kilometers, with storms that rage for centuries and jet streams that encircle the planet at incredible speeds. Yet, for all its power, Jupiter's atmosphere also holds clues to the origins of the solar system. The planet's composition, dominated by hydrogen and helium, reflects the primordial material from which the Sun and planets formed. By studying Jupiter's atmosphere, scientists gain insights into the processes that governed the early solar nebula, helping to unravel the story of planetary formation.

Jupiter's atmosphere also raises questions about the nature of gas giants in general. Observations of exoplanets—planets orbiting other stars—reveal that many share features with Jupiter, from their banded clouds to their immense storms. By understanding Jupiter, we can refine our models of these distant worlds, exploring the diversity of planetary systems and the commonalities that connect them to our own.

Philosophically, Jupiter's atmosphere is a reminder of the forces that shape the cosmos on scales far beyond human experience. It is a world of chaos and beauty, where storms the size of continents churn in perpetual motion and clouds of ammonia and water dance to the rhythms of convection and rotation. Jupiter's atmosphere is both a scientific puzzle and a source of wonder, a testament to the complexity of nature and the interconnectedness of the universe.

To descend into Jupiter's atmosphere would be to enter a realm unlike any other—a journey into crushing pressures, searing heat, and dynamic forces that defy comprehension. It is a place that exists beyond the limits of human survival, yet it holds the key to understanding not only Jupiter but also the processes that shape planetary environments across the cosmos.

Jupiter's atmosphere is more than a feature of the planet; it is a system of staggering scale and intricacy, a living expression of the forces that govern the solar system. In its depths lie answers to questions about the origins of planets, the nature of gas giants, and the dynamics of weather on worlds beyond our own. It is a realm of endless fascination, a reminder of the power and complexity that define the universe.

Chapter 7: Jupiter in Myth and Science

Jupiter has always been a planet of superlatives, and its significance in human history extends far beyond its physical characteristics. Long before telescopes revealed its bands, storms, and moons, Jupiter was a celestial constant, a bright wanderer in the night sky that inspired awe and reverence. In myth, it became a symbol of power, authority, and cosmic order. In science, it transformed into a gateway to understanding the universe. Jupiter's dual identity as both a mythological figure and a scientific marvel underscores its enduring influence on human thought and imagination.

In ancient Roman mythology, Jupiter was the king of the gods, the wielder of thunderbolts, and the guardian of law and order. His Greek counterpart, Zeus, was equally majestic, presiding over Mount Olympus and commanding the heavens. The planet's association with these deities reflects its prominence in the night sky. As the brightest object visible after the Sun, Moon, and Venus, Jupiter naturally became a symbol of dominance and authority. Its steady, unchanging brilliance embodied the idea of cosmic stability, while its imposing presence suggested divine power.

Other cultures also recognized Jupiter's significance. In Babylonian astronomy, Jupiter was associated with Marduk, the chief god of their pantheon, and was seen as a protector of order and justice. Ancient Chinese astronomers referred to it as the "Wood Star," associating it with growth and renewal in their system of five elements. The Mayans tracked Jupiter's movements as part of their sophisticated calendar systems, linking it to cycles of time and celestial harmony.

Jupiter's connection to myth persisted for millennia, even as scientific inquiry began to uncover its true nature. The turning point came in 1610, when Galileo Galilei aimed his telescope at Jupiter and discovered its four largest moons: Io, Europa, Ganymede, and Callisto. This was a revolutionary moment, as Galileo's observations challenged the geocentric model of the universe, which held that all celestial bodies revolved around the Earth. The sight of moons orbiting Jupiter provided clear evidence that not everything in the cosmos was centered on Earth, lending support to the Copernican model of a heliocentric solar system.

Galileo's discovery was not merely a triumph of observation; it was a paradigm shift that redefined humanity's place in the cosmos. Jupiter became a symbol of the vast, intricate universe that lay beyond Earth, a universe governed by natural laws rather than divine fiat. The planet's moons, now known as the Galilean moons, became a focus of scientific study and a reminder of the power of empirical observation.

In the centuries that followed, Jupiter's role in science grew ever more profound. Advances in telescopic technology revealed its bands, its storms, and its faint ring system, while spacecraft missions like Pioneer, Voyager, and Galileo transformed our understanding of the planet and its environment. Each discovery added new layers of complexity to our picture of Jupiter, revealing it as a dynamic system of swirling clouds, powerful magnetic fields, and diverse moons.

Jupiter's significance extended beyond its physical properties. It became a natural laboratory for exploring fundamental questions about planetary formation, dynamics, and evolution. As the largest planet in the solar system, Jupiter holds the key to understanding how planets form and interact with their environments. Its immense gravity has shaped the orbits of countless asteroids and comets, while its composition—dominated by hydrogen and helium—offers a glimpse into the primordial material from which the Sun and planets were born.

The study of Jupiter has also provided insights into the nature of gas giants across the universe. Observations of exoplanets have revealed many worlds similar to Jupiter, both in size and composition. By understanding Jupiter, we refine our models of these distant planets, exploring how they form, migrate, and influence their own planetary systems. Jupiter is not just a planet; it is a template, a reference point for understanding the diversity of worlds that populate the galaxy.

Yet, for all its scientific significance, Jupiter retains its symbolic power. Its storms and auroras remind us of nature's grandeur and ferocity, while its moons—each unique and fascinating—represent the diversity of the solar system. Europa, with its hidden ocean, inspires dreams of

Jupiter

extraterrestrial life, while Io's volcanic activity challenges our understanding of geological processes. Ganymede, with its magnetic field, and Callisto, with its ancient surface, add further layers to Jupiter's story, making the planet a microcosm of cosmic complexity.

Philosophically, Jupiter occupies a unique space in human thought. It is both a beacon of stability and a reminder of chaos. Its steady light has guided travelers and inspired myths, yet its atmosphere is a churning maelstrom of storms and winds. Its gravitational influence protects Earth from potential impacts, yet it can also send comets and asteroids hurtling toward the inner solar system. Jupiter is a paradox, a planet that embodies both order and unpredictability, permanence and change.

As we continue to explore Jupiter, its influence on culture and science deepens. Missions like Juno, which orbits the planet today, provide unprecedented insights into its interior structure, atmosphere, and magnetosphere. Future missions, such as the European Space Agency's JUICE (Jupiter Icy Moons Explorer), promise to unlock the secrets of its moons, particularly Europa and Ganymede. Each discovery adds to the legacy of a planet that has shaped not only the solar system but also humanity's understanding of its place in the cosmos.

Jupiter's story is one of connection—between myth and science, between the past and the future, and between Earth and the heavens. It is a reminder of how far we have come in our quest to understand the universe, and how much more there is to discover. From the ancient myths that saw Jupiter as a god to the modern science that reveals its inner workings, the planet continues to inspire awe and curiosity.

Jupiter is not just a king among planets; it is a bridge between worlds, a symbol of the boundless complexity of the cosmos, and a testament to humanity's enduring drive to explore, learn, and dream.

Chapter 8: Living Under a Giant's Shadow

To live under the influence of Jupiter is to exist in the gravitational and magnetic embrace of a colossus. For billions of years, Jupiter has shaped the solar system with its immense presence, its gravity pulling and redirecting comets, asteroids, and moons, while its magnetosphere dominates the space around it. For the objects and worlds that orbit the giant, including its many moons, life under Jupiter's shadow is a dynamic interplay of protection, challenge, and transformation.

Jupiter is often regarded as the solar system's guardian, its vast gravity acting as a shield for the inner planets, deflecting comets and asteroids that might otherwise collide with Earth or its neighbors. Yet this role as protector comes with an inherent duality. The same gravitational force that sweeps potential threats away from the inner solar system can also hurl objects inward, turning harmless wanderers into dangerous projectiles. Living under Jupiter's influence is to coexist with a force of both preservation and peril.

The moons of Jupiter experience this duality in profound ways. Io, Europa, Ganymede, and Callisto—the Galilean moons—are the most prominent examples, each profoundly shaped by their proximity to the gas giant. Jupiter's tidal forces stretch and flex these moons, creating internal heat that drives volcanic eruptions on Io and may sustain subsurface oceans on Europa and Ganymede. While these forces create environments rich with geological activity and potential habitability, they also expose these moons to intense radiation and the ceaseless tug-of-war of gravitational stresses.

Io, the closest of the Galilean moons, exists in a constant state of upheaval. Its surface is a hellish landscape of erupting volcanoes and sulfuric plains, its interior kept molten by the relentless tidal flexing caused by Jupiter's gravity and the competing pulls of Europa and Ganymede. Io's orbit takes it through Jupiter's magnetic field, generating powerful electric currents that flow back to the planet and contribute to its auroras. Living under Jupiter's shadow has made Io a world of extremes, where the forces that create also destroy, and where change is the only constant.

Europa, farther from Jupiter, experiences a different kind of relationship. Beneath its icy surface lies a vast ocean, kept liquid by the heat generated from tidal interactions. This ocean, shielded from radiation and insulated by kilometers of ice, is one of the most promising locations in the solar system for the search for extraterrestrial life. Yet Europa's existence under Jupiter's shadow is not without cost. The moon is bombarded by intense radiation from Jupiter's magnetosphere, its surface sterilized by charged particles. Living under the giant's influence means Europa's potential as a cradle of life is confined to the dark depths beneath its ice.

Ganymede, the largest moon in the solar system, offers another perspective on life under Jupiter's shadow. Unlike its siblings, Ganymede has its own magnetic field, a feature that creates a complex interaction with Jupiter's magnetosphere. This field provides some measure of protection, reducing the radiation that bombards its surface, while the tidal forces from Jupiter generate internal heating that may sustain a deep subsurface ocean. Ganymede's size and structure make it a microcosm of planetary processes, a moon that mirrors the balance of protection and challenge that defines existence under Jupiter's sway.

Callisto, the outermost of the Galilean moons, exists on the periphery of Jupiter's shadow. Farther from the planet's intense tidal forces, Callisto's surface is a quiet, cratered expanse, a relic of an ancient past. Yet even here, Jupiter's influence is felt. The moon's orbit places it within the fringes of Jupiter's magnetosphere, exposing it to radiation, while its history is intertwined with the gravitational interactions that have shaped the Jovian system.

Jupiter's magnetic field, the largest structure in the solar system, creates another layer of complexity for its moons. The charged particles trapped within this magnetosphere create radiation belts so intense that they pose significant challenges for spacecraft and potential human exploration. These radiation levels are lethal, necessitating advanced shielding and careful mission planning for any probes or habitats that venture near the planet. For the moons that orbit within this field,

Jupiter

radiation shapes their surfaces and environments, adding another dimension to life under Jupiter's influence.

Philosophically, living under Jupiter's shadow invites reflection on the nature of power and balance. Jupiter is both a protector and a force of chaos, a stabilizer and a disruptor. Its presence ensures the order of the solar system, yet it also creates conditions of extremes for those closest to it. The moons of Jupiter are examples of how proximity to immense power can drive creativity and destruction in equal measure.

Jupiter's influence also serves as a metaphor for the interconnectedness of the cosmos. The orbits of its moons, the flow of its magnetic field, and the pathways of comets and asteroids all speak to a system where every element is linked. The balance of forces—gravitational, magnetic, and tidal—creates a dynamic environment where worlds evolve in response to the giant's presence.

For humanity, the idea of living under Jupiter's shadow is not merely a scientific or philosophical concept but a practical challenge. Future missions to explore its moons will need to navigate the radiation belts, manage the tidal forces, and harness the resources of these distant worlds. Europa, in particular, holds the promise of profound discovery, but unlocking its secrets will require overcoming the barriers imposed by Jupiter's immense influence.

Living under Jupiter's shadow is a reminder of the balance between opportunity and challenge, stability and change. It is a reflection of the dynamics that shape not only the Jovian system but the universe itself. To exist in the presence of such a giant is to be shaped by forces beyond comprehension, yet it is also to witness the power of creation and transformation on a cosmic scale.

Conclusion: Jupiter's Dual Role

Jupiter's place in the solar system is one of profound duality—a planet of immense power that both safeguards and endangers, stabilizes and disrupts. It is a celestial giant whose gravitational and magnetic influences extend far beyond its immediate surroundings, shaping the orbits of asteroids, moons, and even the paths of planets. This dual role as protector and potential destroyer encapsulates the paradox of Jupiter, a world whose presence is essential to the order of the solar system but also a force of dynamic unpredictability.

As a protector, Jupiter's gravitational might is unmatched. Over billions of years, it has acted as a cosmic vacuum cleaner, sweeping up debris and deflecting comets and asteroids that might otherwise collide with the inner planets. Without Jupiter, Earth's history would likely have been marked by far more frequent and catastrophic impacts, potentially stunting the development of life or extinguishing it altogether. The scars visible on Jupiter's surface and its moons—left by countless collisions—bear witness to the scale of its protective role.

One of the most striking examples of Jupiter's shield-like function is the impact of **Comet Shoemaker-Levy 9** in 1994. The comet, torn apart by Jupiter's tidal forces, slammed into the planet in a series of dramatic collisions that left visible scars in its atmosphere. These impacts, observed from Earth with unprecedented clarity, highlighted Jupiter's role as a guardian of the inner solar system. Yet they also underscored the power of the forces at play—forces that, under different circumstances, could redirect similar objects toward Earth instead of away from it.

This duality is not confined to celestial impacts. Jupiter's moons live under the constant influence of its gravitational and magnetic fields, which create environments of both challenge and opportunity. Io's extreme volcanism, Europa's hidden ocean, Ganymede's magnetic field, and Callisto's ancient surface all tell stories of worlds transformed by Jupiter's immense presence. These moons, shaped by tidal forces and radiation, illustrate how proximity to a giant can drive both destruction and the potential for new forms of habitability.

Jupiter

Jupiter's magnetosphere, the largest structure in the solar system, is another facet of this dual role. On one hand, it shields its moons from the solar wind, creating a bubble of space dominated by Jupiter's influence. On the other, it generates intense radiation belts that render some regions near the planet incredibly hostile, both to life and to human exploration. The same magnetic field that powers auroras and protects the Jovian system also bombards its moons with charged particles, sterilizing surfaces and creating environments that challenge even the most resilient of life forms.

The planet's atmosphere, with its iconic bands and storms, embodies this tension between order and chaos. The Great Red Spot, a storm that has raged for centuries, is both a symbol of Jupiter's permanence and a reminder of its unpredictable dynamism. The jet streams that define the planet's banded structure are remarkably stable, yet they create vortices and turbulence that reflect the restless energy of the planet. Jupiter's atmosphere is a system of extremes, a microcosm of the balance between stability and change that defines its role in the solar system.

Philosophically, Jupiter's dual role challenges us to think about the nature of power and its consequences. It is a force that ensures the solar system's order, yet it is also capable of introducing chaos. This paradox mirrors the complexities of nature itself, where creation and destruction are often intertwined. Jupiter reminds us that the forces that protect and sustain life can also endanger it, and that existence is shaped by the interplay of stability and disruption.

Jupiter also serves as a metaphor for humanity's relationship with the cosmos. Its gravitational reach connects it to countless objects, from its moons to distant comets, illustrating the interconnectedness of the solar system. In studying Jupiter, we come to understand that no planet or moon exists in isolation; each is part of a larger, dynamic system. This perspective invites us to see our own planet not as an isolated sphere but as part of a broader cosmic story.

The exploration of Jupiter and its moons has deepened our understanding of the solar system's complexity and beauty. Missions like Galileo, Juno, and the upcoming JUICE (Jupiter Icy Moons Explorer) have revealed a planet that is both familiar and alien, a world of swirling clouds, hidden oceans, and magnetic forces that defy easy explanation. Each discovery about Jupiter sheds light on the processes that govern not only our own solar system but planetary systems across the universe.

As we look to the future, Jupiter remains a focal point for exploration and reflection. Its moons, particularly Europa and Ganymede, hold the promise of profound discoveries about habitability and the potential for life beyond Earth. Its magnetic field and atmosphere challenge us to develop new technologies and approaches for studying extreme environments. And its gravitational influence continues to remind us of the delicate balance that allows life to thrive on Earth.

Jupiter's dual role as both savior and potential destroyer is a reflection of the universe's inherent complexity. It is a planet that embodies both the order that sustains and the chaos that transforms, a giant whose influence shapes the destiny of countless worlds. In its vastness and power, Jupiter is a reminder of the forces that govern the cosmos and the delicate balance that makes our existence possible.

To understand Jupiter is to confront the scale of the universe and our place within it. It is a planet that challenges our imagination, inspires our exploration, and reflects the paradoxes of nature itself. Jupiter, the king of the heavens, is both a guardian and a force of transformation—a celestial titan whose dual role continues to shape the story of the solar system.

End Note: The Legacy of Giants

Jupiter is not merely a planet; it is an enduring testament to the grandeur and complexity of the cosmos. Its immense size and influence have earned it the title of "king of the heavens," and its presence has shaped the solar system in ways both subtle and profound. Yet, Jupiter's story is not confined to its towering storms, vibrant auroras, or intricate system of moons. It extends into the lessons it imparts about planetary formation, cosmic interconnectedness, and the forces that govern the universe.

As the largest planet in the solar system, Jupiter provides a window into the earliest stages of planetary formation. Its composition—primarily hydrogen and helium—mirrors that of the Sun, suggesting it formed early in the history of the solar system, capturing much of the material left over from the Sun's creation. In this way, Jupiter serves as a living relic of the primordial solar nebula, a natural archive that holds clues about the conditions under which the planets emerged.

Jupiter's influence extends far beyond its own orbit. Its gravitational pull has shaped the asteroid belt, prevented the formation of another planet between Mars and itself, and guided the trajectories of countless comets and asteroids. It has acted as a stabilizing force, protecting the inner planets from frequent impacts while also catalyzing some of the very events that have shaped Earth's history. Without Jupiter, the solar system as we know it—and perhaps life on Earth—might never have come to be.

Yet, Jupiter's legacy is not merely scientific; it is also symbolic. For millennia, humans have looked to Jupiter as a symbol of power, authority, and cosmic order. It has inspired myths, art, and philosophy, serving as a bridge between the celestial and the terrestrial. In Roman mythology, Jupiter was the king of the gods, a figure of dominance and stability. In modern times, the planet continues to evoke awe and curiosity, embodying the mystery and majesty of the cosmos.

The exploration of Jupiter and its moons has deepened our understanding of the solar system's diversity and complexity. Missions like Galileo and Juno have revealed a planet of swirling storms, dynamic weather systems,

Jupiter

and magnetic forces of staggering power. They have shown us moons with volcanic eruptions, hidden oceans, and ancient surfaces, each telling a unique story of geological and environmental evolution. These discoveries have expanded our understanding of what worlds can be, challenging assumptions and inspiring new questions about habitability and the potential for life beyond Earth.

Jupiter's legacy also extends to the search for exoplanets. Many of the gas giants discovered around other stars bear similarities to Jupiter, from their size and composition to their influence on surrounding planetary systems. By studying Jupiter, we gain insights into the formation and behavior of these distant worlds, connecting our solar system to a broader cosmic context.

Philosophically, Jupiter is a reminder of the balance between creation and destruction, order and chaos. Its immense gravity stabilizes orbits and shields planets, yet it can also redirect comets and asteroids on trajectories of potential devastation. Its storms rage for centuries, yet its presence ensures a measure of order in the solar system. Jupiter is a paradox, a planet of dualities that mirrors the complexities of the universe itself.

As humanity looks to the future, Jupiter remains a cornerstone of exploration and discovery. Its moons, particularly Europa and Ganymede, are among the most promising locations for finding evidence of extraterrestrial life. The technologies and insights gained from studying Jupiter and its environment will inform our approach to other planets and systems, shaping the next chapters of space exploration.

Jupiter's legacy is one of connection—between the past and the future, between Earth and the heavens, and between science and philosophy. It is a reminder of the beauty and power of the universe, a planet that challenges us to think beyond the familiar and to explore the unknown. As we continue to study Jupiter, we are not just uncovering the secrets of a distant world; we are deepening our understanding of the forces that shape existence itself.

Jupiter, the giant among giants, stands as a symbol of the universe's boundless complexity and wonder. Its legacy is a story of discovery,

Jupiter

inspiration, and the enduring human drive to seek out and understand the cosmos.

www.ingramcontent.com/pod-product-compliance
Lightning Source LLC
Chambersburg PA
CBHW070944220526
45469CB00007B/2512